P9-DEG-962

HORNED DINOSAURS

BY "DINO" DON LESSEM
ILLUSTRATIONS BY JOHN BINDON

LERNER PUBLICATIONS COMPANY / MINNEAPOLIS

To Professor John McIntosh, a pioneer in the study of giant dinosaurs

Text copyright © 2005 by Dino Don, Inc.
Illustrations copyright © 2005 by John Bindon
Photographs courtesy of: © Royal Tyrrell Museum/Alberta Community Development, p. 27; © Denis Finnin, American Museum of Natural History, p. 29.
Reprinted in 2006

All rights reserved. International copyright secured. No part of this book may be reproduced, stored in a retrieval system, or transmitted in any form or by any means—electronic, mechanical, photocopying, recording, or otherwise—without the prior written permission of Lerner Publishing Group, except for the inclusion of brief quotations in an acknowledged review.

This book is available in two editions:
Library binding by Lerner Publications Company,
a division of Lerner Publishing Group
Soft cover by First Avenue Editions,
an imprint of Lerner Publishing Group
241 First Avenue North
Minneapolis, MN 55401 U.S.A.

Website address: www.lernerbooks.com

Library of Congress Cataloging-in-Publication-Data

Lessem, Don.
Horned dinosaurs / by Don Lessem ; illustrations by John Bindon.
p. cm. — (Meet the dinosaurs)
Includes index.
Summary: Describes how *Triceratops*, *Torosaurus*, and other horned dinosaurs lived, as well as how scientists have learned about them through fossil studies.
ISBN-13: 978-0-8225-1370-4 (lib. bdg. : alk. paper)
ISBN-10: 0-8225-1370-6 (lib. bdg. : alk. paper)
ISBN-13: 978-0-8225-2574-5 (pbk. : alk. paper)
ISBN-10: 0-8225-2574-7 (pbk. : alk. paper)
1. Ceratopsidae—Juvenile literature. [1. Ceratopsians. 2. Dinosaurs.] I. Bindon, John, ill. II. Title.
QE862.O65L49 2005
567.915—dc21 2003000396

Printed in China
2 3 4 5 6 7 - DP - 11 10 09 08 07 06

TABLE OF CONTENTS

MEET THE HORNED DINOSAURS

WELCOME, DINOSAUR FANS!

I'm called "Dino" Don because I love all kinds of dinosaurs. You've probably heard of some horned dinosaurs, such as *Triceratops.* But what about *Torosaurus* or *Styracosaurus?* Here are the fast facts on some of the most amazing horned dinosaurs. I hope you'll have fun meeting them all.

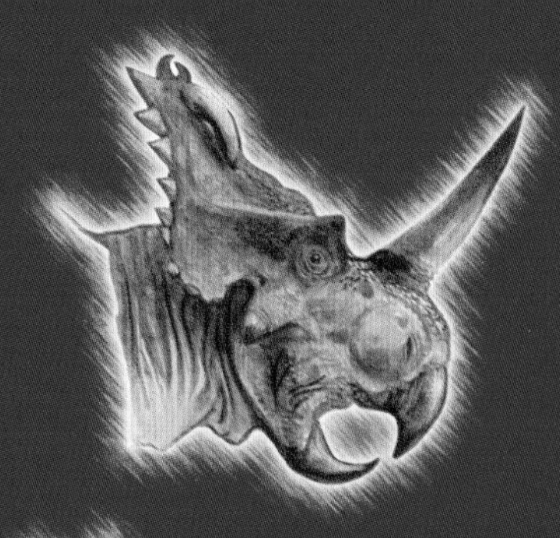

***CENTROSAURUS* (SEN-troh-SAWR-uhs)**
Meaning of name: "sharp-pointed lizard"
Length: 17 feet
Home: northwestern North America
Time: 72 million years ago

***PENTACERATOPS* (PEHNT-uh-SAYR-uh-tahps)**
Meaning of name: "five-horned face"
Length: 28 feet
Home: southwestern North America
Time: 70 million years ago

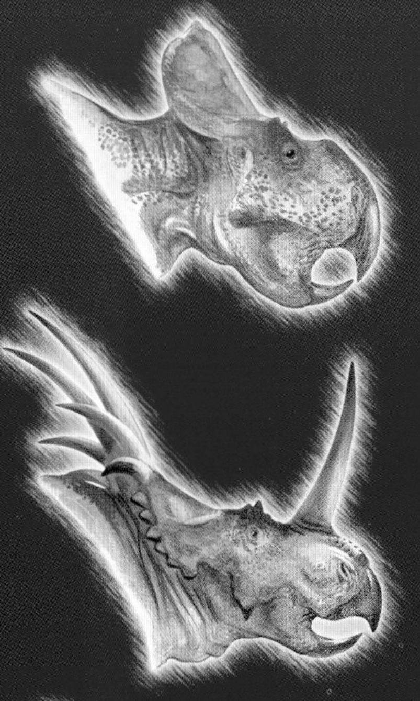

PROTOCERATOPS **(PROH-toh-SAYR-uh-tahps)**
Meaning of name: "first horned face"
Length: 8 feet
Home: eastern Asia
Time: 80 million years ago

STYRACOSAURUS **(STY-rak-uh-SAWR-uhs)**
Meaning of name: "spear-spiked lizard"
Length: 18 feet
Home: western North America
Time: 73 million years ago

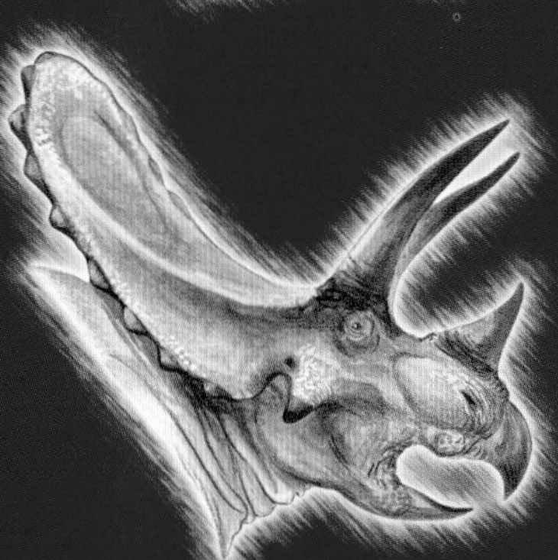

TOROSAURUS **(TOHR-uh-SAWR-uhs)**
Meaning of name: "pierced lizard"
Length: at least 25 feet
Home: western North America
Time: 68 million years ago

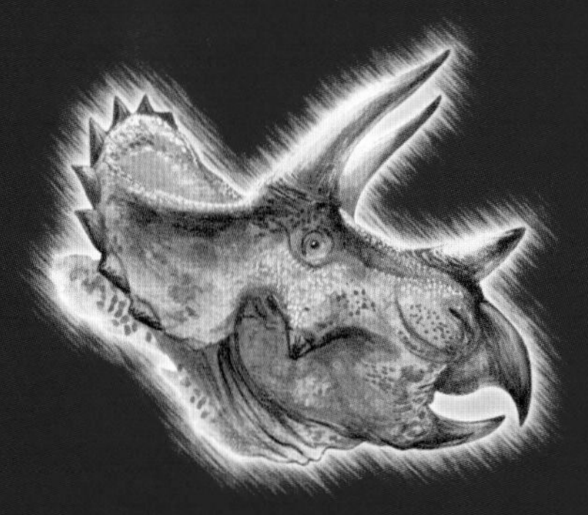

TRICERATOPS **(try-SAYR-uh-tahps)**
Meaning of name: "three-horned face"
Length: 29 feet
Home: western North America
Time: 65 million years ago

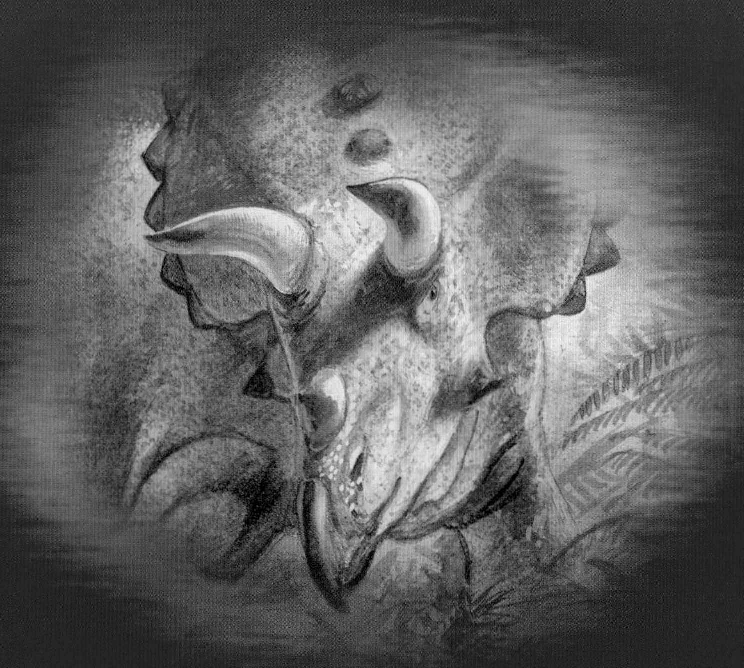

WHAT IS A HORNED DINOSAUR?

BANG! SCRAPE! Noises echo through the forest. Two male *Styracosaurus* are fighting. Each of these horned dinosaurs is as big as an ice-cream truck. They scrape horns and crash against each other's huge skulls.

The *Styracosaurus* aren't trying to kill each other. They're battling to show off for a female. Soon one will back off. The other will be the winner.

THE TIME OF THE HORNED DINOSAURS

Styracosaurus and other dinosaurs lived on land millions of years ago. Scientists used to think that dinosaurs were reptiles. Some dinosaurs had scaly skin, like lizards and other reptiles do. Dinosaurs aren't reptiles, though. Birds are closer relatives of dinosaurs than reptiles are.

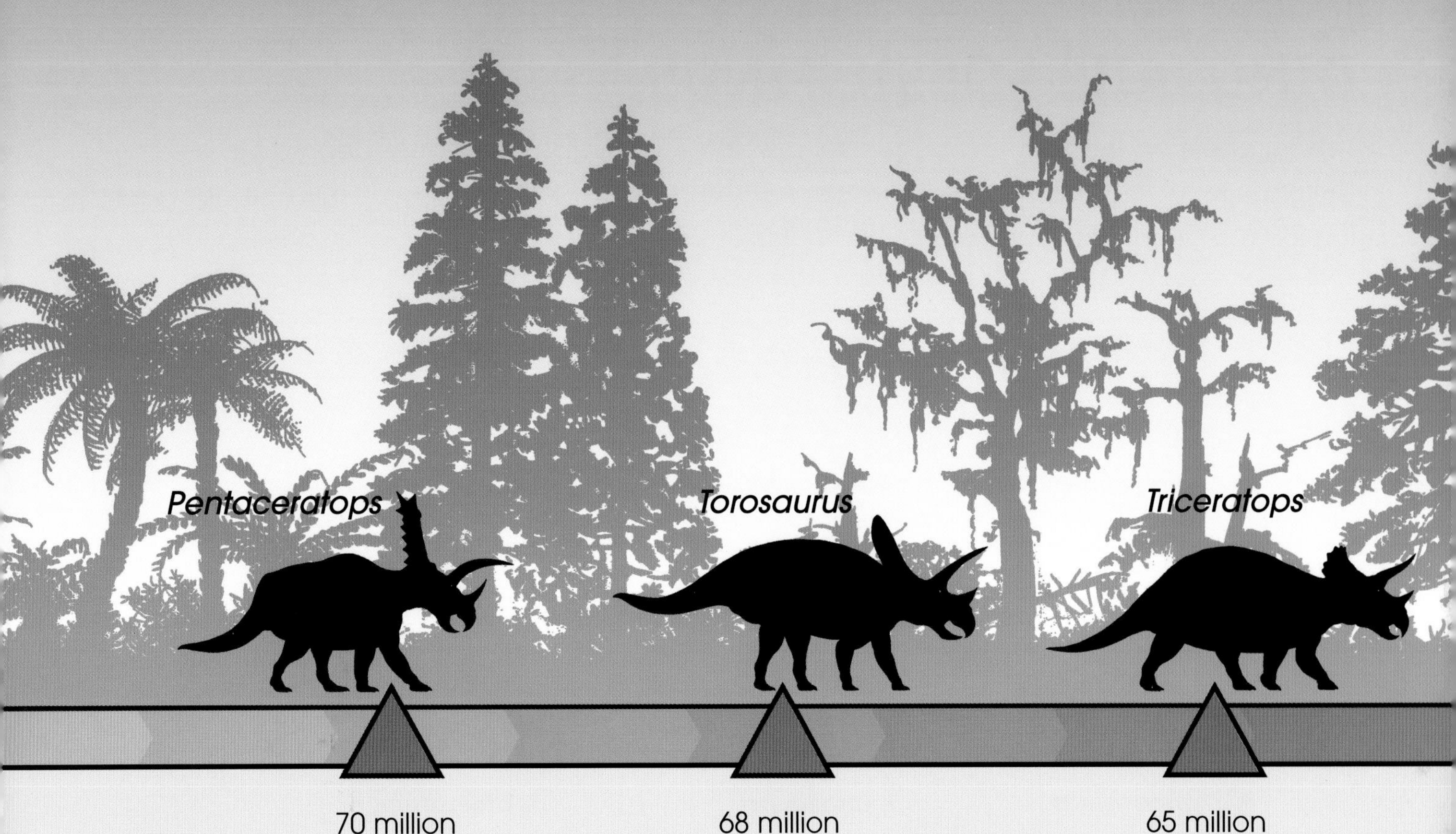

Many kinds of dinosaurs had small horns. But the group we call horned dinosaurs had the biggest horns of all. These dinosaurs were **herbivores,** animals that eat plants. Horned dinosaurs ate with a bony beak called a **rostrum.**

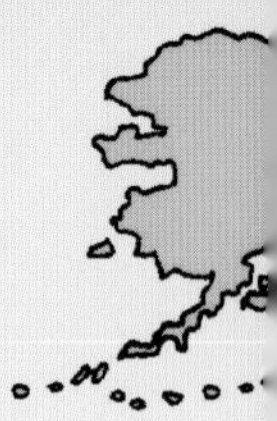

DINOSAUR FOSSIL FINDS

The numbers on the map on page 11 show some of the places where people have found fossils of the dinosaurs in this book. You can match each number on the map to the name and picture of the dinosaurs on this page.

1. Centrosaurus

2. Pentaceratops

3. Protoceratops

4. Styracosaurus

5. Torosaurus

6. Triceratops

We know about horned dinosaurs from the traces they left behind, called **fossils.** Bones and footprints help scientists understand how horned dinosaurs were built. But fossils can't tell us what color a dinosaur was or how its skin looked.

Dinosaurs lived in many parts of the world. But people have found horned dinosaur fossils only in North America and Asia. As far as we know, horned dinosaurs lived only in these places.

HORNS AND FRILLS

Back off, *Tyrannosaurus rex!* *Triceratops* is the largest horned dinosaur. It's almost as long as a school bus! This beast is showing off its horns to scare away the hungry *T. rex.*

The sharp, bony horns of *Triceratops* were longer than a sword. They were probably too thin to hurt a *T. rex.* badly. But they may have scared away *T. rex* and other **predators,** animals that hunt other animals for food.

Look at that huge frill! This ridge of bone grew from the back of a horned dinosaur's skull. Bumps and horns cover the frill of this male *Pentaceratops.* He shakes it so that the female will admire it and choose him as a mate.

A thick, bony frill may have protected a dinosaur from attack. But the frill was probably most useful in attracting a mate. The frill and head were covered with a material like the material in a bird's beak. They may have been brightly colored, like the beaks of some birds.

Two *Torosaurus* shake their big heads back and forth. They're trying to scare each other. Each dinosaur wants to become the leader of his **herd,** a group that lives and roams together.

Torosaurus may have been even bigger than *Triceratops.* Scientists haven't found a complete *Torosaurus* skeleton. But we know that its skull was as long as a whole tiger. That makes it the biggest skull of any animal that ever walked the earth!

GROWING UP

These baby *Triceratops* have just hatched from their eggs. Their parents bring crushed plants to the nest. The adults spit this soft food into the babies' mouths.

Scientists have never found fossils of baby horned dinosaurs in a nest. But all dinosaurs probably hatched from eggs, like birds do. Baby dinosaurs may have been fed in the nest too.

The young *Triceratops* join their herd as soon as they are big enough to walk. The herd searches together for herbs and other plants that grow near the ground. Each dinosaur slices its food with small teeth and a sharp, beak-shaped rostrum.

The young *Triceratops* grow larger horns as time passes. Males probably grew larger horns than females. Males needed big horns to attract mates.

One day, a pair of fierce *Albertosaurus* attacks the herd. The adult *Triceratops* form a circle, their huge heads facing outward. The predators pace, looking for a safe place to attack. Finally, they give up. The herd moves on.

Did horned dinosaurs really form circles to keep their herds safe? We don't know for sure. Some modern horned animals do.

HORNED DINOSAUR DISCOVERIES

In 1923, scientists found a nest of dinosaur egg fossils. It was near the bones of many *Protoceratops.* So the scientists thought the eggs were *Protoceratops* eggs. A fossil of another dinosaur lay nearby. Was it raiding the nest? The scientists called it *Oviraptor.* This name means "egg thief."

Years later, the bones of an *Oviraptor* were found on top of another fossil nest. The eggs looked the same as those in the first nest. But one held a fossil of a baby *Oviraptor.* The first *Oviraptor* hadn't been stealing horned dinosaur eggs after all. It had been guarding its own nest!

Fossils have also taught us how horned dinosaurs died. This huge herd of *Centrosaurus* was crossing a river when heavy rains caused a flood. The dinosaurs panicked. Some drowned. Others were crushed in the rush for dry land.

How do we know this disaster happened? Scientists found hundreds of *Centrosaurus* fossils buried in rocks in western Canada. The rocks were made of the kind of sand found on river bottoms. So these dinosaurs must have died in a river.

Predators ended the lives of many horned dinosaurs. This *Velociraptor* attacks a *Protoceratops* with claws as sharp as razors. Who will win the battle?

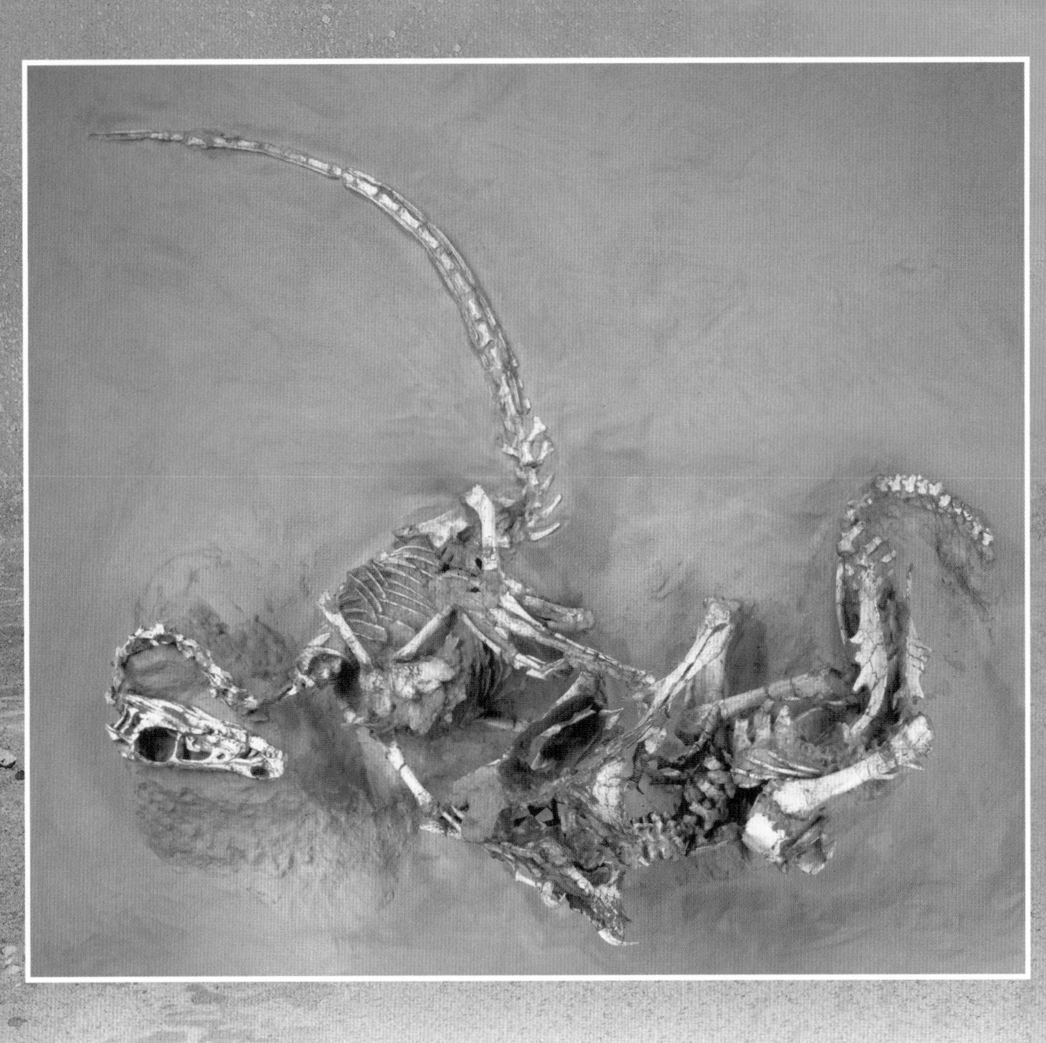

Both dinosaurs lost. Scientists found their skeletons buried in sand. A sandstorm may have covered them as they fought. Or a sandy hill may have fallen on them. Their skeletons slowly turned into rock.

The last horned dinosaurs died out 65 million years ago. Scientists think that's when a huge **asteroid** struck the earth. The asteroid may have sent clouds of dust into the air. The dust would have changed the weather, killing plants and animals.

The asteroid crash may have been one reason that the horned dinosaurs died out. They are gone forever. But they left behind many traces of their lives. These fossils have shown us many things about the lives of the amazing animals we call horned dinosaurs.

GLOSSARY

asteroid (AS-tur-oyd): a large, rocky lump that moves in space

fossils (FAH-suhlz): the bones, tracks, or traces of something that lived long ago

herbivores (URB-uh-vohrz): animals that eat plants

herd (HURD): a group of animals that live, eat, and travel together

predators (PREH-duh-turz): animals that hunt and eat other animals

rostrum (RAHS-truhm): a bony beak

INDEX